Akinola Babatunde
Serkan Abbasoglu

Photovoltaic system performance in countries with high solar radiation

AF 138183

Akinola Babatunde
Serkan Abbasoglu

Photovoltaic system performance in countries with high solar radiation

A case study of Northern Cyprus

LAP LAMBERT Academic Publishing

Impressum / Imprint

Bibliografische Information der Deutschen Nationalbibliothek: Die Deutsche Nationalbibliothek verzeichnet diese Publikation in der Deutschen Nationalbibliografie; detaillierte bibliografische Daten sind im Internet über http://dnb.d-nb.de abrufbar.

Alle in diesem Buch genannten Marken und Produktnamen unterliegen warenzeichen-, marken- oder patentrechtlichem Schutz bzw. sind Warenzeichen oder eingetragene Warenzeichen der jeweiligen Inhaber. Die Wiedergabe von Marken, Produktnamen, Gebrauchsnamen, Handelsnamen, Warenbezeichnungen u.s.w. in diesem Werk berechtigt auch ohne besondere Kennzeichnung nicht zu der Annahme, dass solche Namen im Sinne der Warenzeichen- und Markenschutzgesetzgebung als frei zu betrachten wären und daher von jedermann benutzt werden dürften.

Bibliographic information published by the Deutsche Nationalbibliothek: The Deutsche Nationalbibliothek lists this publication in the Deutsche Nationalbibliografie; detailed bibliographic data are available in the Internet at http://dnb.d-nb.de.

Any brand names and product names mentioned in this book are subject to trademark, brand or patent protection and are trademarks or registered trademarks of their respective holders. The use of brand names, product names, common names, trade names, product descriptions etc. even without a particular marking in this work is in no way to be construed to mean that such names may be regarded as unrestricted in respect of trademark and brand protection legislation and could thus be used by anyone.

Coverbild / Cover image: www.ingimage.com

Verlag / Publisher:
LAP LAMBERT Academic Publishing
ist ein Imprint der / is a trademark of
OmniScriptum GmbH & Co. KG
Heinrich-Böcking-Str. 6-8, 66121 Saarbrücken, Deutschland / Germany
Email: info@lap-publishing.com

Herstellung: siehe letzte Seite /
Printed at: see last page
ISBN: 978-3-659-80688-9

ACKNOWLEDGEMENT

All gratitude goes to the almighty God who kept me alive and well to pursue this work.

I want to thank my Mother/Guardian, Justice M.A Folayan and every member of my family who supported me and made provisions available to complete this work.

Thanks to my wife, Abisola Babatunde, whose patient love, prayers, encouragement and attention enabled me give my best to this study. Thank you for sustaining the hope and vision that kept me going.

To my supervisor, who is also a Co-Author of this work, Assoc. Prof. Dr. Serkan Abbasoglu; your assistance, encouragement and invigorating recommendations in the time of our research and documentation of this work are appreciated.

My appreciation goes to Cyprus International University Library for ensuring that both hard copies and online resource materials relevant to the successful completion of this research were made available.

Special thanks to Cyprus Solar Technology, Nicosia whose PV systems were the real world applications considered in this study. Your assistance and cooperation helped in improving the value of this research.

TABLE OF CONTENT

ACRONYMS AND ABBREVIATIONS

AC	Alternating current
AOI	Angle of incidence
ARC	Anti-reflection coating
a-Si	Amorphous silicon
BIPV	Building integrated photovoltaics
BOS	Balance of system
CdTe	Cadmium Telluride
CIS	Copper indium di-selenide
c-Si	Crystalline silicon
CF	Capacity factor
CYS	Cyprus solar technology
DC	Direct current
DOE	Department of energy
E_p	Energy production
EU	European union
Ey	Specific yield
FF	Fill factor
GIS	Geographic information system
IEA	International energy agency
I_{mpp}	Current at maximum power point
I_{sc}	Short circuit current
NOCT	Nominal operating cell temperature
NC	Northern Cyprus
NPO	Nominal plant output
NREL	National renewable energy laboratory
P_{max}	Maximum power
PR	Performance ratio
PV	Photovoltaics
PVT	Photovoltaic thermal
STC	Standard test conditions
V_{mpp}	Voltage at maximum power point
V_{oc}	Open circuit voltage
W_p	Peak watt

LIST OF TABLES

LIST OF FIGURES

CHAPTER 1
INTRODUCTION

1.1   ENERGY

Energy can be considered as one of the key components and constituents of a society. It is usually required to provide various services. Economic development and improved standard of living depend on adequate and reliable supply of energy. Energy is a basic concept in all sciences and engineering disciplines. It is well known that energy cannot be created or destroyed but just converted and redistributed from one form to another; such as from solar energy into electrical energy, or chemical energy to mechanical energy. Our current living standards could not be maintained without energy. Total world energy demand and consumption by sources and country are shown in the Figure 1.1 and 1.2 below. This trend indicates that demand for energy will continue to increase.

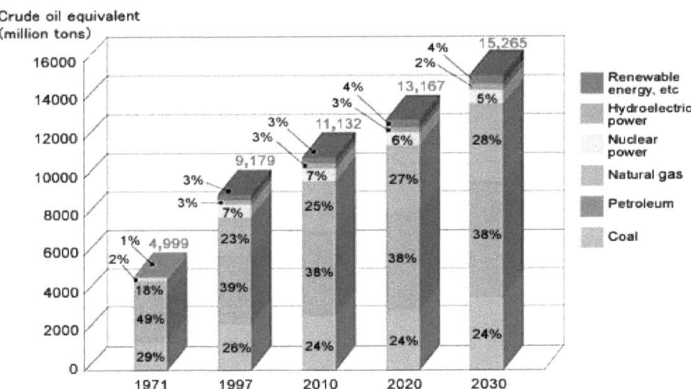

Figure 1.1: Trends and forecast of world energy demand by fuel

(Source: Agency for natural resources and energy, Japan)

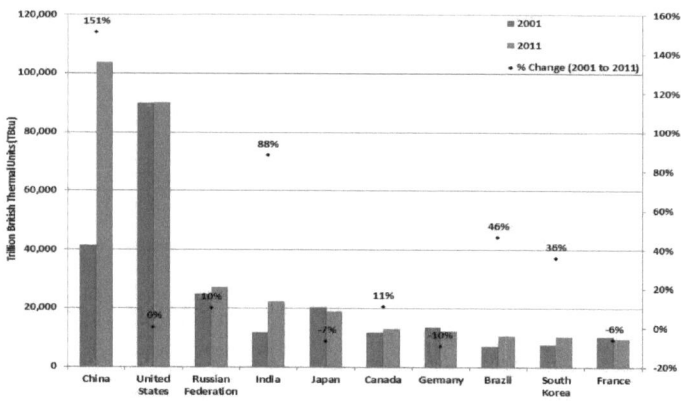

Figure 1.2: World energy demand by country
(Source: BP statistical review of world energy, June 2012)

1.1.1 Renewable Energy

Energy sources can be classified into non-renewable and renewable sources. There are different types of renewable energy sources; solar energy, wind energy, bioenergy, tidal energy and geothermal energy. Renewable energy sources can be defined as the sources that are not depleted with continuous use, are environmentally friendly, and are not hazardous to health. The worries about the sustainability of conventional energy sources (fossil and nuclear fuels) are the main advantages of the renewable energy sources. Concerns about the adverse environmental and social consequences of fossil fuel use, such as air pollution or mining accidents, and about the finite nature of supplies, have been voiced intermittently for several centuries. In addition, the price rise of the fossil fuels due to the oil crisis and the advent of environmental movement, force the humanity to begin to take seriously to find an alternative to these fuels. (Boyle, 2004.)

1.1.2 Solar Energy

This is the energy from sunlight. Due to the abundance of sunlight, solar energy is said to be the most promising renewable energy for the future. Solar energy is void of

2

carbon emissions; hence it has little or no hazardous effect on the environment and human health. Solar radiation can be converted into useful energy directly by using various technologies. When absorbed in solar collectors, it can provide hot water, or space heating. Buildings can also be designed with passive solar features that improve the contribution of solar energy to space heating and lighting requirements. Mirrors can be used to concentrate solar energy to provide high temperature heat for generating electricity as solar thermal electric power stations for commercial operations. Solar radiation can also be converted directly into electricity using photovoltaic (PV) modules. Nicosia, Northern Cyprus (NC), is our main focus of this study. Nicosia is known to have abundance of sunlight; it has an average of 300 sunny days in a year, its average daily sunny hours is 11 and average annual solar radiation is 5.2 kW/m² daily. (Abbasoglu S., 2010). This positions NC as a fertile location for conversion of solar energy to electricity.

1.2 PHOTOVOLTAIC SYSTEM

"Photovoltaic PV system is an elegant solar energy source. Light shines on a crystal and produces electricity. There are no moving parts. The fuel source (sunlight) is free, abundant and widely distributed, available to every country in the world. At over 165,000 TW the solar resource dwarfs the world's current power usage of 16 TW or even the projected future usage of 60 TW. PV is a solar energy technology that makes use of characteristics of certain semiconductors to directly convert solar radiation to direct current (DC) electricity. PV systems utilize wafers, which are made of crystalline silicon; the crystalline silicon responds to sunlight and produces a small direct current when they come in contact with light. These PV cells or solar cells can be combined into larger sized arrangements called modules, the modules arranged or combined together systematically are referred to as arrays; they produce large appreciable amount of electrical power with no moving parts and noise, and minimized emissions. A PV system can be referred to as an electrical system consisting of a PV array and other electrical components needed to convert solar energy into usable electricity by various types of loads and/or appliances. The components can be arranged in many ways to design PV systems for different situations, but the most common configuration is the utility connected system, which is found on commercial and residential buildings" (Dunlop P. 2010.)

1.2.1 Advantages of PV systems

PV systems have several advantages when compared to other energy sources. Some of these advantages are:

1. PV is an environmentally friendly technology that produces energy with no noise or pollution. Therefore operating a PV system makes a statement about protecting the environment and conserving non-renewable energy sources.
2. PV systems are flexible and can be adapted to many different applications. The modular nature of PV arrays and other components make systems easy to expand for increased capacity.
3. There are no moving parts, hence PV system are extremely reliable and last a long time with minimal maintenance.
4. PV systems offer energy independence. A grid connected PV system reduces the consumer's vulnerability to utility power outages and lifelong bills, and a stand-alone system eliminates it.
5. Sunlight, which is the energy source for PV systems, is free and readily available.
6. PV system costs are generally decreasing as conventionally produced electricity is expected to become more expensive, hence PV systems can be used to hedge against future energy price increases.

1.2.2 Disadvantages of PV systems

Despite several advantages, PV systems also have some demerits as listed below:

1. PV systems have high initial cost compared to competing power generating technologies.
2. They require relatively large array area to produce significant amount of power.
3. The available solar radiation resource at a particular location determines the feasibility of producing appreciable amount of power.
4. Limited or lack of knowledge about the potential of solar PV systems.

1.2.3 History and Development of PV Systems

The term photovoltaic is derived by combining the Greek word for light, photo, with volt, the name of the unit of electromotive force – the force that enables the motion of electrons (i.e. electric current). The volt was named after an Italian physicist Count Alessandro Volta, the inventor of the battery. PV thus describes the generation of electricity directly from light. The French physicist Edmund Becquerel discovered photovoltaic effect. In 1839, he published a paper describing his experiments with a wet cell battery, in the course of which he found that the battery voltage increased when its silver plates were exposed to sunlight (Dunlop P. 2010.)

In 1877, two Cambridge scientists, W.G Adams and R.E Day, described the variations they observed in the electrical properties of selenium when exposed to light (Dunlop P. 2010.) Charles Edgar Fritts, a New York electrician, in 1883, constructed a selenium solar cell that was similar to the silicon solar cells of today. But, his cell was very inefficient. The efficiency of a solar cell is defined as the percentage of the solar energy falling on its surface that is converted to electrical energy. Less than 1% of the solar energy falling on these early cells was converted to electricity. Selenium cells eventually came into widespread use in photographic exposure meters (Boyle, 2004.) In 1948, two

Bell Telephone Laboratories (Bell Labs) researchers, Bardeen and Brattain, produced revolutionary device using semiconductors - the transistor. Transistors are made from semi-conductors (usually silicon) in extremely pure crystalline form, into which tiny quantities of carefully selected impurities, such boron and phosphorus have been deliberately diffused. This process, known as doping, dramatically alters the electrical behavior of the semiconductor in a very useful manner.

It was not until the 1950s that the breakthrough occurred that set in motion the development of modern, high-efficiency solar cells. It took place at the Bell Labs in New Jersey, USA, where a number of scientists, including Darryl Chapin, Calvin Fuller and Gerald Pearson, were researching the effects of light on semiconductors. In 1953, Chapin, Fuller and Pearson produced doped silicon slices that were much more efficient than earlier devices in producing electricity from light. By the following year there had been successes in increasing the conversion efficiency of their silicon solar cells to 6%.

1.2.4 Global Installed Capacity of PV Systems

The total PV capacity worldwide as at early 2013 is a little above 100 GW. The solar industry is on track to install another 100 GW worldwide by 2015, nearly doubling solar capacity in the next 2.5 years (Stephen, L. 2013). The installed capacity of PV systems is shown in Figure 1.3.

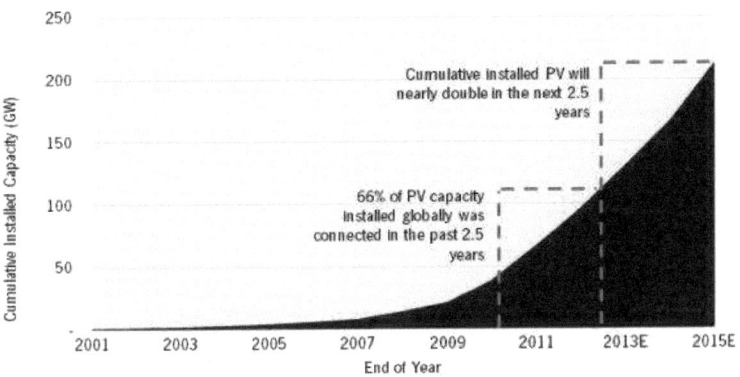

Figure 1.3: Cumulative global installations at year end (2001 – 2015E)

(Source: GTM Research)

Globally, 31.1 GW solar PV capacity was installed in 2012; this is an all time annual high that pushed global PV capacity above 100 GW. While PV production has become increasingly concentrated in one country, China, the number of countries installing PV is growing rapidly. In 2006, only a handful of countries could boast of solar capacity of 100 megawatts or more. Now 30 countries are on that list, which the International Energy Agency (IEA) projects will more than double by 2018. Today roughly 60% of PV is manufactured in China (see Figure 1.4). A decade ago, China produced almost no PV. But in a kind of gold rush spurred by easy bank loans and government tax incentives and subsidies, China hurtled past PV technology pioneers - the United States (in 2006) and Japan (in 2008) as shown in Figure 1.4.

Worldwide, PV production in 2012 declined 2 percent from 2011 (Figure 1.5), the first annual drop on record. But, this contraction will be short-lived as demand continues to rise. Solar power installations are growing more than 40 percent annually, and falling PV prices are making solar power more affordable (Roney J.M. 2013).

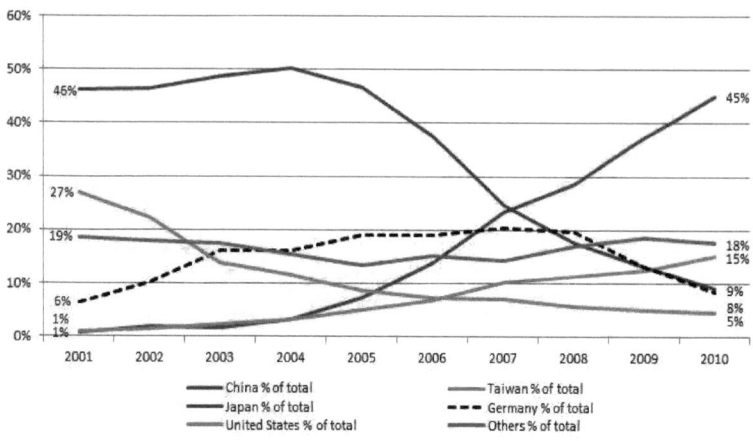

Figure 1.4: Largest producers of solar PV, % of total world production (2001 -2010)

(Source: Earth Policy Institute)

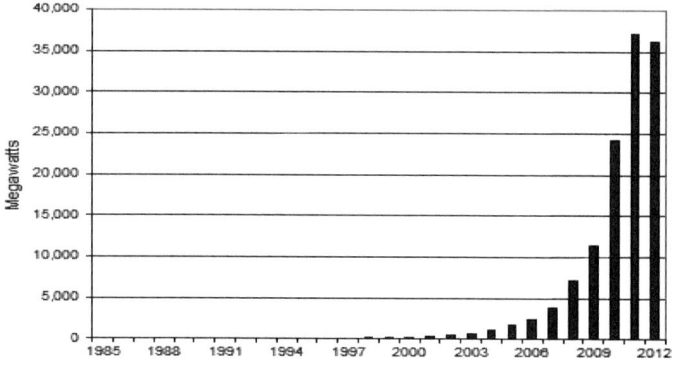

Figure 1.5: World annual solar PV production

(Source: PV news; GTM research)

1.2.5 Global Prices of PV Systems

Global module prices have fallen 62% since January 2011 (Stephen, L. 2013). Figure 1.6 shows that price of silicon crystalline PV dropped substantially in the past thirty years. As the installed cost of solar PV drops to $4/watt, the price per kilowatt-hour (kWh) (depending on climate and geography), will be equal to about $0.16/kWh that would be in line with utility rates after rates caps are removed. It is seen that despite the lower PV panel costs; PV technology is still not at parity with hydrocarbon fuels such as coal and oil. Carbon based taxes or renewable energy incentives as well as more investment into alternative energy should improve the economics of solar and wind and lead to grid parity (Anonymous, 2009). US DOE confirms that the current module cost for solar PV varies from $5.5 to $10 per watt. According to NREL, the price of electricity for solar PV ranges from $0.15 to $0.59 per kWh.

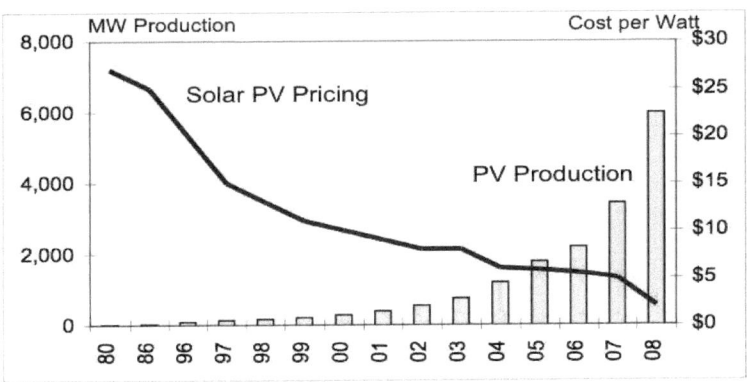

Figure 1.6: Solar global production and cost per watt
(Solar Buzz, Company Reports, Green Econometric Research)

1.3 OBJECTIVE OF THIS STUDY

The main focus of this work is summarized in three parts as follows:

1. Measurement and analysis of single axis tracking PV system field data with respect to the meteorological conditions of Nicosia. These field parameters include energy production, specific yield, performance ratio, and capacity factor.

9

2. Validation and comparison of the PV system's field measurements with literature and also three different simulation results; and providing detailed explanation of variance and similarities in these results.

3. Prediction of best case PV system performance using the PV simulation software tools.

1.4   WORK ORGANIZATION

This work is divided into three main chapters. Chapter 1 contains the literature review, the critical points of current knowledge including substantive findings, as well as theoretical and methodological contributions to field measurements of PV systems globally and especially in Mediterranean weather conditions were highlighted. Examinations of the key points and concepts in previous studies were reviewed.

Chapter 2 focused on the methods, theories, and techniques used in this study. The meteorological conditions of Nicosia which include the annual average solar radiation and duration ambient temperature etc. were highlighted. The characteristics of the PV power plant deployed for field measurements were described. PV software tools utilized for this research were presented and their limitations and advantages explained. The sources of various data inputs, assumptions made and the rationale behind these assumptions were detailed.

Results and discussions were presented in Chapter 3. The performance of CYS PV system was discussed in detail. Various factors responsible for variation in parameters for different seasons of the year were analyzed and their effects were discussed. Comparison of CYS PV system with the PV simulation tools was also presented.

CHAPTER 2
LITERATURE REVIEW

Simple PV systems ensures the production of power, not only for small consumer items such as calculators and wristwatches, but also for some complex systems such as communications satellites, water pumps, home and workplace appliances such as lights, machines etc. A lot of bill/advertisement boards, road and traffic signs also now get powered by solar PV. A PV system consist of many solar cells; a single PV/solar cell is usually small; it produces approximately 1 or 2 watts of power. The PV cells are arranged and interconnected together in series or parallel to form larger units referred to as modules in order to increase the output power of PV cells, The modules can also be connected to form much larger units called arrays, which can be interconnected to generate more power. With this in mind, PV systems can be installed to meet mostly any small or large electric power need. Several other structures and components are coupled with modules or arrays to build a whole PV system. Some structures point them toward the sun, and there are also components that take the DC electricity produced by modules and condition that electricity, mostly by converting it to alternate-current (AC) electricity. Batteries may be incorporated into some PV systems. All of these items are known as the *balance of system* (BOS) components. Combining modules with BOS components create an entire PV system. PV system components are classified as shown in Table 2.1. Figure 2.1 shows a fixed grid connected PV system with various components.

Table 2.1: PV system components (Dunlop P., 2010)

Classification	Components
Modules and Arrays	Primary components common to all PV systems
Energy Storage Systems	Batteries, Flywheels, Super-capacitors
Power Conditioning Equipments	Inverters, Charge Controllers, Rectifiers and Chargers, DC-DC Converters, Maximum Power Point Trackers.
Electrical Loads	DC loads (lighting fixtures, motors for fans and pumps), AC load (refrigerators, conditioners, television, lightings and motors)
Balance of System (BOS)	Electrical BOS (conductors, cables, conduits, junction boxes, enclosures, connectors, and terminations needed to make circuit connections between modules, controllers, batteries, inverters) Mechanical BOS (fasteners, brackets, enclosures, racks and other structural supports for safety and reliability)

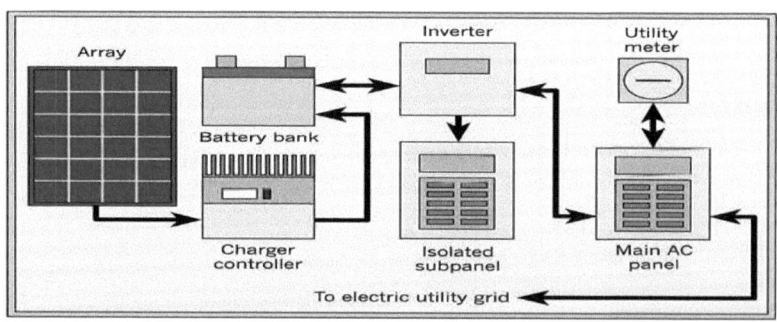

Figure 2.1: PV components and connection (Source: ecmweb.)

2.1 PV SYSTEM CONFIGURATION

The optimal PV configuration for any application depends on load usage, load type, solar resource, auxiliary power sources and other factors (Dunlop P., 2010). A large number of configurations/applications are possible for PV systems; this is summarized in the Table 2.2. Figures 2.2-2.4 show simple block diagrams for each of these configurations.

Table 2.2: PV system configurations

PV Configurations	Characteristics
Stand-Alone Systems	Operate autonomously and supply power to electrical loads independently of the electric utility. They store energy in batteries. Used when other energy sources are impossible considering costs. Classifications include: Direct-Coupled Systems, Self-Regulated Systems, and Charge-Controlled Systems.
Grid-Connected Systems (also known as Utility-Interactive Systems)	Operate in parallel with and is connected to utility grid. They are the simplest and least expensive PV systems that produce AC power as they require the fewest components and do not use batteries. Primary component is the inverter. These systems are modular and scalable.
Grid-Connected Systems with Battery (also known as Bimodal Systems)	The inverter is the main component; it draws DC power from the battery system instead of the array. The array simply acts as charging source for the battery system. These systems operate in a manner similar to UPS.
Hybrid Systems	These are stand alone systems that include two or

more distributed energy sources such as PV arrays, engine generators, wind turbines, and micro-hydroelectric turbines. Hybrid systems offer the advantage of greater system reliability and flexibility in meeting variable loads. Using a variety of energy sources may also reduce total system costs. They are the most complex of all PV systems in terms of equipment, system design, and installation. Classifications include DC and AC bus hybrid system.

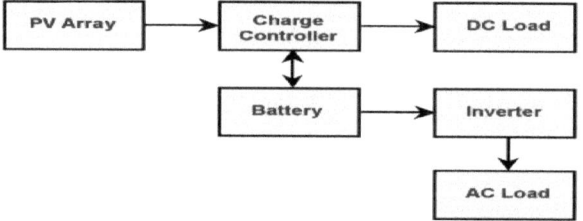

Figure 2.2: Stand-Alone PV system with battery

(Source: Florida Solar Energy Center)

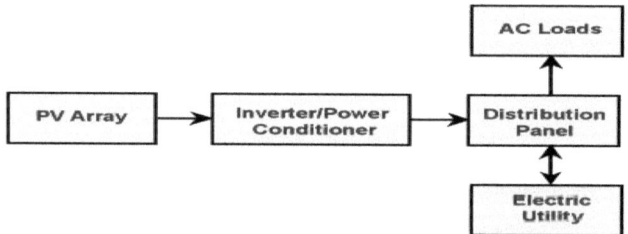

Figure 2.3: Grid-Connected PV system with battery

(Source: Florida Solar Energy Center)

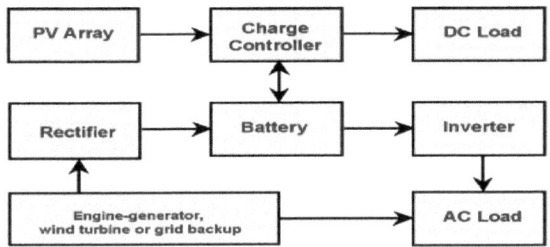

Figure 2.4: Hybrid PV system

(Source: Florida Solar Energy Center)

2.2 PV ARRAY ORIENTATION

PV Array Orientation refers to the direction, position and angle of PV arrays with regards to the sun. As explained in reference (Dunlop P., 2010.), the orientation of PV system increases or decreases how much solar radiation can be accessed by a PV system. Two main angles are used to define array orientation: *tilt angle and azimuth angle*. The *tilt angle* is the vertical angle between the horizontal and array surface. Maximum annual solar energy from arrays on a fixed surface is achieved by orienting the surface at a tilt angle nearly the same as the value of the local latitude. Depending on the energy needs, array can be optimized for either summer or winter gain by changing the tilt angle. The *azimuth* angle is the horizontal angle between a reference direction and the direction an array surface faces. Reference direction is usually due north or south. For fixed mounts, azimuth angle is due south.

In their research with HOMER software, Amita C. et al. (2008) found that the optimum tilt is different for each month or season of the year for various locations. The study concludes that the yearly average of optimum tilt is equal to the latitude of the site. When the tilt angle and azimuth angle are made to experience continuous changes by following the movement or position of the sun, we have what is called sun tracking. Sun tracking may be classified according to the number and orientation of the axes used to

15

track the sun. *Single-axis tracking* is a sun tracking system that rotates one axis to approximately follow the position of the sun. This brings significant performance gains over fixed surfaces. Single-axis tracking is primarily used for flat plate systems, and sometimes with concentrator systems. *Dual-axis tracking* rotates two axes independently to exactly follow the position of the sun. This further maximizes the amount of solar energy received, though the gain over single axis tracking is smaller than the gain of single-axis tracking over fixed surfaces. Dual-axis tracking is primarily used with PV concentrators.

Five common ways of mounting fixed PV systems are: The *Ground Mount Systems* use a framing system similar in design and orientation to standard passive slanted roof mounting systems, except they are mounted on the ground adjacent to the building or source of consumption. These systems are used when the desired electrical output of the system is limited due to a lack of roof mounting area. Soil testing and the use of the appropriate anchoring system for the environmental characteristics of the site are necessary.

Another type is *Building Integrated Photovoltaics (BIPV)*; these are architecturally pleasing mounting systems that can be installed on many building surfaces including vertical walls. Wiring is run through interior walls or the support structure with the associated equipment stored in equipment rooms or basements. *Photovoltaic Solar Canopy, Walkway & Awning Systems* are unique in that each application has its own architectural design, load and secondary use characteristics. Like building integrated photovoltaics, semi-transparent panels are often used and mounted using a process similar to glazing systems used in skylight construction. All the internal wiring is contained within the structural mounting system. In applications where the structure is stand-alone, the balance of the wiring can be run underground or through conduit, while the associated electrical components are housed in the adjacent building.

Pole Mounted Systems – these are essentially ground mounts without the framing system. Solar arrays are attached to the top or side of the mounting pole. These types of mounts can allow for manual or mechanical adjustment of the panel orientation for seasonal tracking of the panels to the overhead sunlight path. *Roof Mounting Systems* are the most commonly seen portrayals of mounted solar panels. They are both affordable and fairly easy to install. Other advantages to roof mounting systems are: protection from vandalism, unwanted & accidental contact; wide variety of substrate attachment options. Height may make the system less affected by shading & obstructions too. Finally, these systems can be customized for a specific application which is another major advantage. These framing systems are generally comprised of aluminum, stainless steel for additional strength and anti-corrosion or a combination of both.

2.3   COLLECTION OF PV DATA

Sharma V. et al. (2013) carried out a predicted and measured performance assessment of different solar photovoltaic technologies under similar outdoor conditions. The authors proposed that the percentage difference between measured and predicted output can be minimized by using more accurate input data to the PV simulation software, rather than using the data available in the simulation software. The study assessed the performance of three different photovoltaic technology modules: polycrystalline silicon, hetero-junction with intrinsic thin-layer silicon and amorphous single junction silicon. It was also observed that, there were high variations after comparing predicted and measured performance of the three different PV technologies.

2.4   PV SYSTEM PERFORMANCE

In order to analyze the performance of a PV system, the performance parameters specified by International Energy Agency (IEA) include: the total energy yield (E_Y), specific yield (Y_A), and performance ratio (PR). These are the important parameters which provide information about the overall system performance, with respect to energy production and system losses (Marion B et al., 2005). According to U.S. Department of Energy, the most common PV array design uses flat-plate PV modules or panels. These panels can be fixed in a particular position or designed to track the movement of the sun. They have the capacity to respond to either direct or diffuse sunlight. When the sky is clear, the diffuse part of sunlight accounts for between 10% and 20% of the total horizontal surface solar radiation. On partly sunny days, up to 50% of that radiation is diffuse, and on cloudy days, 100% of the radiation is diffuse. PV systems use both diffuse and direct radiation for energy production while certain systems such as Concentrator PV systems use only direct radiation for energy production. PV module performance is measured with peak watt ratings. The peak watt (W_p) rating is determined by measuring the maximum power of a PV module under laboratory conditions of relatively high light, favorable air mass, and low cell temperature. But these conditions are not totally obtainable in real world applications. Due to this, researchers normally use a different procedure, known as the normal operating cell temperature (NOCT) rating. In this procedure, the module first equilibrates with a specified ambient temperature so that maximum power is measured at a nominal operating cell temperature. This NOCT rating results in a lower watt value than the peak-watt rating, but it is probably more realistic. PV modules convert solar energy to electricity directly. The amount of energy produced by PV system is dependent on a number of factors. These factors that affect PV systems performance are summarized as follows:

- *The temperature of the PV module* which in turn depends on the ambient (air) temperature, the solar irradiance, the type of mounting, and cooling by wind (Huld T. et al., 2010)

- *The PV module energy conversion efficiency*: This is generally a non-linear function of the irradiance level and module temperature, typically declining for low irradiances and for high temperatures (Huld T. et al., 2010). For some module types the conversion efficiency depends on the history of the module. Efficiency changes with exposure to light (negatively for some module types, such as amorphous silicon (a-Si), positively for other types, such as Copper–Indium–Diselenide (CIS) and Cadmium Telluride (CdTe). Such an effect is often termed light-soaking. Exposure to elevated temperatures may also affect the conversion efficiency, in particular in the annealing effect in a-Si modules which acts to counteract the light-soaking (delCueto and von Roedern, 1999; Nikolaeva-Dimitrova et al., 2008).

- *The fraction of sunlight reflected away at the module surface*: This depends on module type and on the incidence angle of the light relative to the surface. This is termed the angle-of-incidence (AOI) effect. This effect lowers the irradiance entering the generating layers in the solar cells of the modules and hence affects both temperature and power output (Martin and Ruiz, 2001; Huld T. et al., 2008).
- *Module types*: They vary in their spectral sensitivity. Instantaneous solar spectra in turn depends on the meteorological conditions in a way that is not well understood, especially for cloudy conditions (Gottschalg et al., 2004, 2005).

2.5 SOLAR IRRADIATION

When the light intensity incident on a solar cell is varied, virtually all the solar cell parameters change. These parameters are short circuit current; open-circuit voltage; fill factor (FF); efficiency and the influence of series and shunt resistances. The light intensity on a solar cell is called the number of suns, where 1 sun corresponds to standard illumination at AM1.5, or 1 kW/m^2. For example a system with 10 kW/m^2 incident on the solar cell would be operating at 10 suns, or at 10X. A PV module designed to operate under 1 sun conditions is called a "flat plate" module while those using concentrated sunlight are called concentrators.

Solar irradiance is the power of solar radiation (energy that emanates from a source in form of waves or particles) per unit area (kW/m^2). It is used as a reference condition to evaluate the output performance of a solar energy system at a given point in time or for rating the power output of solar energy utilization equipment such as PV modules. *Solar irradiation* (kWh/m^2), therefore, is the total amount of energy accumulated on an area over time. Solar irradiation is the principal data needed for sizing and estimating the performance of PV systems. Greater solar irradiance (power) means energy is accumulated faster, which result in greater solar irradiation (total energy). Solar irradiance (solar power) is measured with a *pyranometer*. A *pyranometer* measures *total global (direct and diffuse)* solar irradiance in a hemispherical field of view. Diffuse global radiation can be measured by shading a pyranometer from the direct radiation component. The direct radiation measurement is then calculated by subtracting the diffuse measurement from global measurement. Another way of calculation direct solar radiation is to use a *pyrheliometer*. This is a sensor that measures only direct solar radiation in the field of view of the solar disk (5.7°); it does not measure diffuse radiation component (Dunlop P., 2010).

In Makrides G. et al. (2010) study, they point out that PV modules installed in Cyprus are subject to high irradiation and high module temperature reaching 60°C in the summer which favors their performance. The annual solar irradiation measured on site in Cyprus is 1997 kWh/m^2 with the maximum contribution during the summer period. They also pointed out that lower annual irradiation for the same time period was measured by the same type of pyranometer in Stuttgart (Germany) and shows annual irradiation of $1460kWh/m^2$; this is in line with the study and documentations of Abbasoglu S. et al., (2010) that NC has higher solar radiation compare to Germany (world's solar market leader).

2.6 TEMPERATURE OF PV MODULE

In the study of Swapnil Dubey et al. (2013), PV modules with less sensitivity to temperature are preferable for the high temperature regions and those more responsive to temperature will be more effective in the low temperature regions. He further explains that solar cell performance decreases with increasing temperature, fundamentally owing to increased internal carrier recombination rates, caused by increased carrier concentrations. It is of importance to note that the electrical performance of PV systems is primarily influenced by the type of PV used. A typical PV module converts 6-20% of the incident solar radiation into electricity, depending upon the type of solar cells and climatic conditions. The rest of the incident solar radiation is converted into heat, which significantly increases the temperature of the PV module and reduces the PV efficiency of the module. This heat can be extracted by flowing water/air beneath the PV module using thermal collector, called, photovoltaic thermal (PVT) collectors. In practice, only amorphous silicon (a-Si) and crystalline silicon (c-Si) have been found in the literature on PVT. The higher efficiency of c-Si will result in a higher electrical efficiency and a higher electrical-to-thermal ratio of the PVT than in the case of a-Si.

The experiment of Griffith J. S et al (1981), confirms that the operating temperature plays a central role in the photovoltaic conversion process; it also shows that both the electrical efficiency and the power output of a PV module depend linearly on the operating temperature decreasing with PV cell temperature. Makrides G et al. (2010) evaluated the temperature behavior of different photovoltaic systems installed in Nicosia-Cyprus and Stuttgart-Germany. The results were compared with manufacturer's data and reasonable agreements were achieved. First two technologies, multi-crystalline (also known as polycrystalline) and thin film were compared in Cyprus, it was observed that multi-crystalline show larger maximum power point voltage (V_{mpp}) and open circuit voltage (V_{oc}) temperature coefficients. Also the peak powers (P_{mpp}) for mono-crystalline and multi-crystalline were larger than the respective coefficients of thin film technologies. In another research and publication by Makrides G. et al. (2008), it was seen that most installed fixed flat plate PV systems in Cyprus produce annual ac energy yield between 1600-1700 kWh/kWp. For the same period of time, the tracker produced 30% higher ac energy yield over the average fixed plate energy yield for the same period.

2.7 ELECTRICAL CHARACTERISTICS OF PV MODULES

Solar cells can be connected in different ways. We have series connection and parallel connection. In series connection, all the voltages add up. The current, however remains constant; this means that the current flowing through a single cell is the same as the current flowing through a string of cells. Strings of cells refer to many cells connected together. In parallel connection, all the currents in a string of cells add up while the voltage remains constant irrespective of the number of cells. The resulting voltage at the output of a module is called open-circuit voltage (V_{oc}) and the resulting current is called short circuit current (I_{sc}). At the point where the solar cell produces maximum power (P_{max}), the voltage and current at such point are referred to as maximum power point voltage (V_{mpp}) and maximum power point current (I_{mpp}). Figure 2.5 represents the current-voltage (I-V) and power curve of a solar cell.

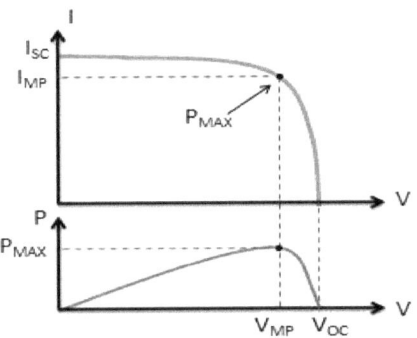

Figure 2.5: I-V curve and power curve of solar cells

(Source: Samlexsolar)

Bypass diodes are sometimes incorporated into solar modules. When we look at solar modules in real life, there are possibilities of it being partly shaded. This shading could be as a result of neighboring buildings or nearby trees etc. Shading can have adverse effect on the output of a solar module. If a cell in a solar module is shaded, it means current flowing through the cell will be reduced. In a series connection the current is limited by the cell producing the lowest current. This cell dictates the maximum current flowing through the module. If a constant load is connected to the module, there will be a voltage drop across the load due to lower current generated. The other non-shaded cells will be forced to produce more voltage and act as a reverse bias source to the shaded cell. Meaning, no energy is generated in the shaded cell but energy is dissipated. This makes the solar cell to start getting warmer. The temperature can get to a critical level whereby the encapsulation of solar material cracks or other materials begin to wear out. Hence high temperature leads to decrease in PV output. This situation can be prevented by incorporating a bypass diode in the module. The use of bypass diode can help in reducing the undesired effects of a dysfunctional solar cell. The bypass diode is connected in parallel to a solar cell with a polarity opposite to that of the solar cell. If no

23

cell is shaded or dysfunctional, no current will flow through the diode. Figure 2.6 and 2.7 shows the connection of a bypass diode to a module and characteristic graph respectively.

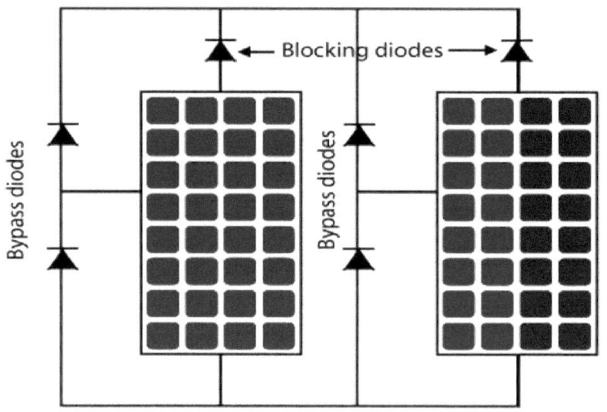

Figure 2.6 By-pass diode PV connections

(Source: sunwindproduct)

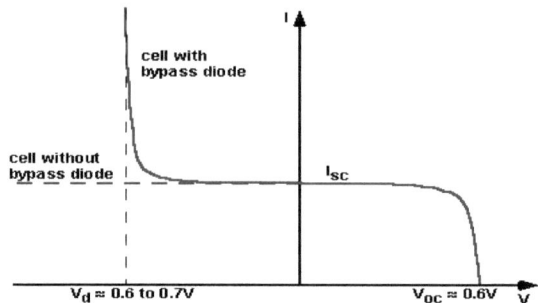

Figure 2.7: By-pass diode PV curve

(Source: PVEducation)

24

The electrical behavior of solar cells is also affected by temperature and irradiance. This effect of temperature is shown in Figure 2.8.

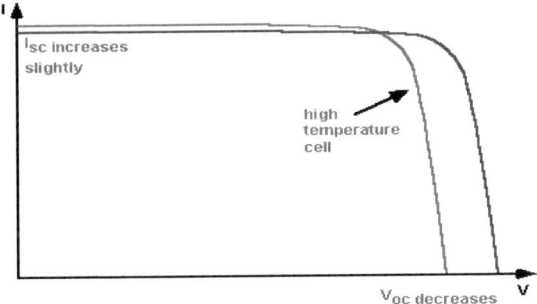

Figure 2.8: Effect of temperature on solar cells

(Source: PV Education)

Solar cells are very sensitive to variation in temperature. An increase in temperature decreases the band gap of a semiconductor. When the band gap of a semiconductor is decreased as a result of increase in temperature, the energy of the electrons in the material is increased. This means lower energy will be required to break the bond. In a solar cell, the parameter most affected by an increase in temperature is the open-circuit voltage. Power produced by PV module relies on solar irradiation value. The open circuit voltage and maximum power voltage are not influenced greatly by the amount of irradiance but the short circuit current and the maximum power current relies directly on the solar irradiation hitting the surface of the PV module (Figure 2.9).

25

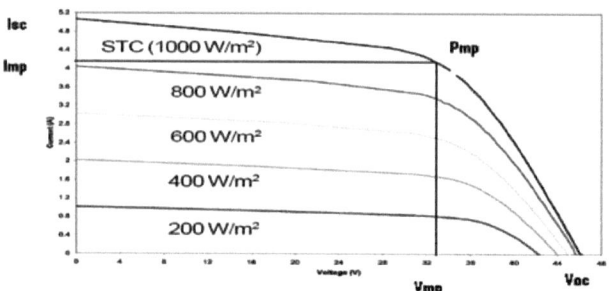

Figure 2.9: Effect of solar irradiation on solar cells

(Source: Century Suntech)

Generally, PV technology is classified according to their generations. We have the first (1G), second (2G) and third (3G) generation PV as shown in Figure 2.10. 90% of global PV installation is based on the first generation technology of crystalline silicon wafers such as the monocrystalline and polyscrystalline silicon wafers. The thin film silicon PV is the second generation technology; they are cheaper but have not been able to match the efficiency of first generation cells. Their lattice is either amorphous or nanocrystalline.

	1G	2G	3G
Technology	Wafer-based (single junction, mono- or polycrystalline silicon)	Thin film (CdTe, CIGS, a-Si)	Organic (Polymer, Small Molecule), DSSC
Advantages	High quality, low-defect, high efficiency	High material utilisation, lower cost	Non-toxic, abundant, low-cost, short payback, transparent
Challenges	High consumption of active material (Si)	Scarcity or toxicity of some materials	Optimizing lifetime-efficiency-cost trade-off

Figure 2.10: First, second and third generation PV technology characteristics

(Source: www.novaled.com)

26

CHAPTER 3
METHODOLOGY

This study evaluates the performance of a 5.76 kW PV system in Nicosia, NC and also compares the measured data from the system with the results obtained from using three different PV simulation software tools. The software tools used for design and simulation are pvPlanner, PVsyst and Homer. First, performance analysis is carried out by the evaluation of four different performance indicators; energy production, performance ratio, specific yield and capacity factor. These parameters will be discussed and defined in this chapter in detail. Then, the results are evaluated and validated with studies in literature. Also, these results are compared with the results of three different PV simulators and the differences are evaluated.

3.1   SYSTEM DESCRIPTION

The system is located in Nicosia. Figures 3.1 (a) and (b) gives a cross section of the 5.76 kW PV power system. It is tilted at 30° on latitude 35.2°N and longitude 33.4°E. A full description of the 5.76 kW is given in Table 3.1.

Figure 3.1(a): 5.76 kW Solar Power Plant at CYS

Figure 3.1(b): 5.76 kW Solar Power Plant at CYS

Table 3.1: CYS PV system description and technical data at STC

Description	Specification	Description	Specification
Manufacturer	Mage Powertech Plus	Tracking	Single – axis
Solar module type	Polycrystalline silicon cell	Module efficiency	15.1%
Nominal power (Pnom)	240W_p	Number of strings in parallel (array)	2
Voltage at Pnom	29.57V	Number of solar modules in series (strings)	12
Current at Pnom	8.2A	Total number of modules	24
Short circuit current (I_{sc})	8.76A	Tilt angle	30°
Open circuit voltage (V_{oc})	37.35V	Weight	20 kg
Inverter	2 x SMA SB3000 HF	Dimensions (L x W x D)	(1670 x 1000 x 50) mm
Standard test conditions (STC)	1000W/m², AM1.5, 25°C	Single module area	1.626m²

The system module conversion efficiency is 15.1%. This indicates that only 151 Wh/m²
of the 1,000 Wh/m² (the standard test condition power density) is converted by the PV

28

module to useful direct current electrical energy. A global summary of some PV module efficiencies is shown in Table 3.2. Lazar R. (2013) defines the operating efficiency of a solar panel as the ratio between electric power delivered to the load and incident light intensity. It is affected by the following four main factors: cells temperature, sunlight intensity, sun angle, and load. This means efficiency depends not only on the properties of PV cells, but also on the environment and the load.

Table 3.2: Ratings chart of solar panels rated above 100 watt with highest efficiency
(Source: Lazar R., 2013)

Manufacturer	Model	Watts	Efficiency (%)	Vmpp (V)	Imp (A)
SunPower	SPR-327NE-WHT-D	327	20.10	54.7	6.0
Crown Renewable Energy	CR100	100	18.30	17.6	5.7
Xinhonglian Solar Energy	XHH135-24	135	18.01	36	3.8
Zhejiang Shuqimeng Energy Tech. Co., Ltd.	SE230M-33A / D	230	17.94	46.7	4.9
Sanyo Electric	HIP-205BA19	205	17.70	56.7	3.6
Shanghai Pubsolar	GYS-280D	280	17.69	36.4	7.7
Sanyo Electric	HIP-205BA20	205	17.66	56.7	3.6
Titan Energy Systems	TITAN 32-300	300	17.24	53.8	5.5
Enfoton Solar Ltd	60E6M+245F(L)	235	17.06	30.1	8.1
Shanghai Pubsolar	GYS-270D (270)	270	17.05	35.7	7.6
Symphony Energy	SE-S160	160	16.95	30.2	5.4
Conergy AG	STM 210 FWS	210	16.88	40	5.3
Millennium Electric T.O.U. Ltd.	MIL-PVT-215W-M03	215	16.84	38.2	5.6

Maaß Regenerative Energien GmbH	BN 180-SP	180	16.53	23.5	7.7
Solara AG	SM 400SP	100	16.52	17.8	5.6
Centrosolar America	SM 400SP	100	16.52	17.8	5.6

3.2 METEOROLOGICAL CONDITIONS OF NICOSIA

Nicosia, the capital of NC, is located on latitude 35.17^0 and longitude 33.36^0 in northern hemisphere as shown in Figure 3.2.

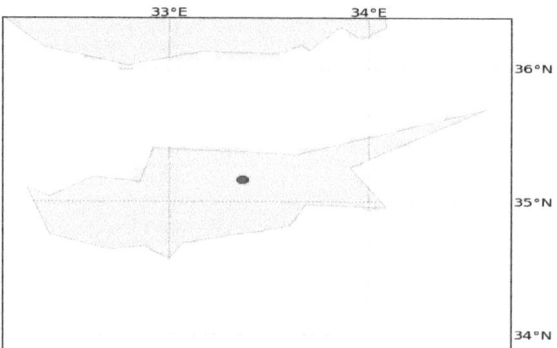

Figure 3.2: Location (longitude and latitude) of Nicosia

(Source: Solar GIS)

Solar irradiation in Cyprus (a typical example of an island in the Mediterranean sunbelt) is one of the highest in Europe, with more than 300 days of the year considered as having sunny weather and an annual irradiation of around 2,000 kWh/m² on a tilted surface of 27.58°, which is much higher than the sunniest areas of the world's largest market, Germany (Zinsser B. et al., 2007). The total land mass of Cyprus is about 9,200 km² with North Cyprus occupying the northern part of the island, with an area of 3,355 km². Figure 3.3 shows annual radiation on optimally inclined surfaces.

30

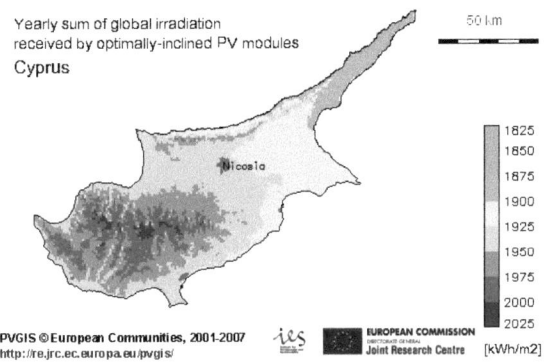

Figure 3.3: NC annual radiation on optimally-inclined surface
(Source: PVGIS)

The climate of Cyprus is a typical Mediterranean climate with hot, dry summers. And during winter, the weather is warm and rainy in the day, becoming cold at night. The average solar radiation received in Cyprus, based on historical data collected by the various meteorological stations on the island is estimated at 7.0 kWh/m^2/day in summer and 3.0 kWh/m^2/day in winter months (Ibrahim D. et al., 2012).

3.3 PV SIMULATION SOFTWARES

This study compares outdoor data with three different PV simulation software tools that are well known globally and accepted for PV simulations. These software tools are pvPlanner, PVsyst and Homer energy. There are several PV simulation software tools across the globe. Some software tools are commercially licensed while some are not. Also, different versions of these software tools exist based on either improvement processes or specific applications. Each tool has its mathematical and atmospheric modeling for irradiance and temperature. The basic characteristics of the simulation tools for this study are summarized in Table 3.3.

Table 3.3: Simulation characteristics

(Source: Sandia National Laboratory and various PV Software Websites)

	pvPanner	**PVSyst**	**Homer**
Version assessed	Not applicable	5.63	2.81
Commercially licensed	Yes	Yes	Yes
Meteo data input	15-minute time series data	1 year of hourly data	1 year of hourly data
Irradiance model (or Plane of array)	Geo-Model Solar	Hay and Davis Model or Perez et al	HDKR model (Hay, Davis, Klucher and Reindl)
Array performance model	-	One diode equivalent circuit model (modeled for thin film)	Linear irradiance model with temperature correction

3.3.1 pvPlanner simulation software

According to SolarGIS website, "pvPlanner is one of the numerous applications of SolarGIS. SolarGIS is a geographical information system designed to meet the needs of solar energy industries. With pvPlanner one can accurately calculate PV electricity potential within minutes. It is the best companion for site prospection. It is also the ideal tool for comparing energy yield from various PV technology options (e.g. crystalline vs amorphous silicon modules) and benefits from different mounting systems (e.g. fixed, 1-axis trackers, 2-axis trackers, etc.). The major advantages of pvPlanner include:

- Integration with SolarGIS database - use only the most accurate solar radiation data for simulations

- User friendly - easy to use not only for engineers but also for business managers
- Assurance from experts - authors of pvPlanner are leading experts in PV simulation.

Simulations methods used in pvPlanner are most advanced and scientifically validated. In pvPlanner, photovoltaic power production is simulated using numerical models developed or implemented by GeoModel using aggregated data based on 15-minute time series of solar radiation and air temperature data as inputs. Data and model quality is checked according to recommendation of IEA SHC Task 36 and EU FP6 project MESoR standards. The simulation itself is quite complex process. The computation process consists of 8 main steps (see fig 3.4). When finished, the SolarGIS server sends the output to user's browser where the data are shown in forms of Tables and graphs. For PV simulation there are two kinds of basic inputs:

- Site parameters: provided by SolarGIS database
- Technical parameters: provided by pvPlanner user (or default values will be taken into consideration)"

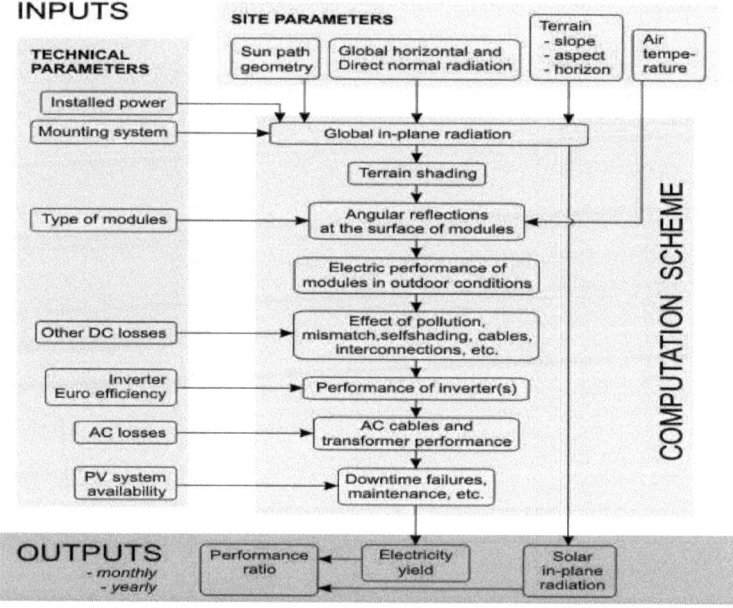

Figure 3.4: pvPlanner computation scheme

3.3.2 PVsyst simulation software

In PVsyst website, "PVsyst is designed to be used by architects, engineer, and researchers. It is also a very useful educative tool. It offers a user-friendly approach with guide to develop a project. PVsyst is able to import meteorological data from many different sources, as well as personal data. PVsyst presents results in the form of a full report, specific graphs and Tables, and data can be exported for use in other software. It is a software package for study, sizing and data analysis of complete PV systems. It deals with grid-connected, stand-alone, pumping and dc grid PV systems. The software offers three levels (preliminary stage, project design and measured data analysis) of PV system study corresponding to different stages of the development of real project." Unlike pvPlanner, PVsyst offers the flexibility of accepting user solar radiation input. It also allows the user the platform to design each component to taste and specification.

34

3.3.2 Homer simulation software

"The HOMER energy modeling software is a powerful tool for designing and analyzing hybrid power systems, which contain a mix of conventional generators, combined heat and power, wind turbines, solar photovoltaics, batteries, fuel cells, hydropower, biomass and other inputs. It is currently used all over the world by tens of thousands of people. HOMER is a micropower optimization model which simplifies the task of evaluating designs of both off-grid and grid-connected power systems for a variety of applications. HOMER's optimization and sensitivity analysis algorithms make it easier to evaluate many possible system configurations. HOMER simulates the operation of a system by making energy balance calculations in each time step of the year. For each time step, HOMER compares the electric and thermal demand in that time step to the energy that the system can supply in that time step, and calculates the flows of energy to and from each component of the system. It then determines whether a configuration is feasible, i.e., whether it can meet the electric demand under the conditions one specifies, and estimates the cost of installing and operating the system over the lifetime of the project. The system cost calculations account for costs such as capital, replacement, operation and maintenance, fuel, and interest." (Paul Gilman, 2004).

"Engineers and non-professionals can use HOMER to run simulations of different energy systems, then compare the results and get a realistic projection of capital and operating expenses. As distributed generation and renewable power projects continue to be the fastest growing segment of the energy industry, HOMER can serve utilities, telecoms, systems integrators, and many other types of project developers - to mitigate the financial risk of their hybrid power projects. It also allows solar radiation input by the user" **(HOMER energy website).**

3.4   THE USE OF DATA INPUT

pvPlanner does not allow user to input solar irradiance, it provides it's irradiance from its data base. Generally the solar irradiance and temperature used for the three simulation software tools (pvPlanner, PVsyst and HOMER) were those obtained from meteorological databases of the software tools. At the time of this study, the readings from the pyranometer connected to the system at site was not available, so we obtained Nicosia solar radiation and temperature for the system specification from the NC meteorological office. The PV system nominal power, PV efficiency, number of modules, inverter size and efficiency, tilt angle and other data obtained (see Table 2.1) from filed measurement were fed into the simulation softwares as inputs, as accurately as possible (within various software limitations).

The selection of solar cell was more accurate for PV syst, as it is the only software that allows for such project precision. Polycrystalline silicon from universal energy, model UE 240-P2 series, which is about the closest to cell/module of the system at site was selected. Homer software provides a generalized PV module without the option of technology selection. This may be one of the reasons that accounted for the overstatement of the output energy as shown in the next chapter. In pvPlanner, which provides more specificity for cell selection than Homer, only three technologies are available: crystalline silicon, amorphous silicon and cdTe; crystalline silicon was chosen but it was not clear it was mono- or poly- crystalline.

Also, the choice of inverter was limited with Homer and pvPlanner. They were only able to accommodate inverter efficiency and percentage losses. Homer accepted the size of inverter but pvPlanner does not have such platform. In PVsyst, an accurate selection of inverter used in the field power plant was possible which is Sunny Boy SB 3000HF manufactured by SMA. The size, efficiency and units were all to precision. This also may be part of the reasons for the accuracy seen in the results obtained from PVsyst simulation tool.

3.5   SOLAR PV PERFORMANCE PARAMETERS

The solar PV performance parameters examined in this study are: energy production, specific yield, performance ratio and capacity factor.

3.5.1   Energy Production

This can be classified into the direct current (DC) energy and the alternating current (AC) energy. This study focuses on the analysis of the AC energy. The electrical energy production of a photovoltaic plant is the total alternating current (AC) energy delivered to the grid in the case of grid-connected systems (without primary load applications). It is calculated by multiplying the operating power of the PV system in kW and duration of operation of the system in hours as shown in the equation below.

$$Energy\ production\ (kWh) = Power\ (kW)\ x\ time\ (hours)$$

The monthly and annual AC energy production of the system studied was measured from the system energy meter, analysis is made on the energy production values for the winter and summer months. This analysis takes into consideration the effect of solar radiation, solar duration and ambient temperature during the period considered. The system energy production is then compared to the energy production simulated by each of the three PV software tools. The similarities and variance were observed and analyzed.

3.5.2   Specific yield

"The specific yield of a PV system is the net energy output divided by the nameplate D.C power at Standard Test Condition (AM1.5, 1000 W/m^2, 25^{0}C) of an installed PV array. It represents the number of hours that the PV array would need to operate at its rated power or peak rating to provide the same energy. The units are hours or kWh/kW_p. The specific yield normalizes the energy produced with respect to the system size," (Marion B. et al., 2005). The specific yield is calculated from the equation below:

$$Specific\ yield = \frac{AC\ energy\ output\ (kWh)}{Nameplate\ or\ DC\ rated\ power\ on\ PV\ module\ (kW_p)}$$

In this study, the specific yield values obtained were used to determine the total hours the PV system operated at its peak rating daily, monthly and annually. The period of the year with lowest and highest peak rating hours were determined. The PV software tools also simulated the monthly and annual specific yield; the simulated yields were then compared with the system specific yields.

3.5.3 Performance ratio

Cristian P. et al., (2009) defines performance ratio (PR) of a given PV system as energy output injected into the grid (AC part) divided by the name plate DC energy obtained under standard test condition of PV array. PR does not have any unit and as explained earlier, it is normally reported on a monthly or yearly basis and its values can also be calculated for smaller intervals, such as weekly or daily. This is useful for identifying occurrences of component failures depending on geographical location and season. Performance ratio is widely considered the best measure of panel quality, because all components and their interactions are taken into consideration during calculation. PR is calculated using the equation below:

$$PR = \frac{Actual\ AC\ energy\ reading\ of\ plant\ output\ (kWh)}{Calculated\ nominal\ plant\ output(kWh)}$$

The actual plant output is read from the export grid meter. Nominal power output (NPO) is calculated as follows:

$$NPO = \frac{Incident\ solar\ radiation\ on\ PV\ system}{Irradiance \times Area \times Efficiency\ of\ the\ PV\ modules}$$

The monthly and annual PR of the PV system were obtained, variation of in the PR at different seasons of the year were analyzed. The values of system PR and the simulated PR values were compared.

3.5.4 Capacity factor

Capacity factor (CF) is sometimes referred to as capacity utilization factor. It does not have any unit and can be calculated from the equation below:

$$CF = \frac{Energy\ measured\ (kWh)}{365\ days \times 24\ hours \times Installed\ capacity\ of\ the\ plant}$$

PR is a measure of the performance of a PV system taking into account environmental factors (temperature, irradiation, etc.) while CF completely ignores all these factors and also the de-rating or degradation of the panels. CF does not take into account the availability of the grid and minimum level of irradiance needed to generate electrical energy. As discussed earlier, PR can be used as a tool to compare different solar PV systems with each other, even if they are installed at different locations since all environmental factors will be taken into account. CF can only compare systems design and the ability of the systems to convert solar energy into electrical energy. The PV system studied does not have a provision for the measurement of CF, the values represented in this study were calculated theoretically using the equation above. Also, the only PV simulation tool that modeled CF is HOMER, the value for other software tools were calculated theoretically.

CHAPTER 4
RESULTS AND DISCUSSIONS

In order to analyze the performance of the 5.76 kW single-axis tracking Cyprus Solar (CYS) PV system located at Nicosia, the following performance parameters will be considered in detail: the *total energy production, specific energy yield, performance ratio (PR)* and *capacity factor (CF)*. These parameters are defined and discussed in the previous chapters. These are some of the important parameters which provide information about the overall system performance (Marion B et al., 2005).

The solar radiation and ambient temperatures and parameters that affect the results directly, are evaluated primarily. The PV simulators modeled their results with slightly different radiation and temperature data from measured data for Nicosia. pvPlanner does not allow for manual input of solar radiation and temperature, hence we did not manually input the measured radiation and temperature into any of the simulators. pvPlanner utilizes solar radiation data from GeoModel solar, calculated from the satellite and atmospheric data from Meteosat PRIME satellite, and it provides 15-minute data values. HOMER solar radiation data is obtained from HDKL model which is also referred to as Hay, Davis, Klucher and Reindl model. PVsyst solar radiation data was obtained from Hays model (1979) and Perez (1987, 1988) model" (Geoffrey T. K et al, 2009. p. 20, 31)

The global-in-plane radiation for CYS, pvPlanner and PVsyst are 2565 kWh/m², 2580 kWh/m² and 2377 kWh/m² respectively. Figure 4.1(a) and (b) shows the annual and monthly global horizontal radiation for Nicosia in comparison with the simulators. It can be seen that, among the PV simulators, pvPlanner has a data set with the highest annual solar radiation (1,892 kWh/m²), followed by HOMER with 1,881 kWh/m². The measured global horizontal solar radiation of Nicosia at the system site is 1,754 kWh/m² in 2012. Also, from the measured data, Nicosia monthly solar radiation peaks at 203 kWh/m² in July and crests at 66 kWh/m² in January.

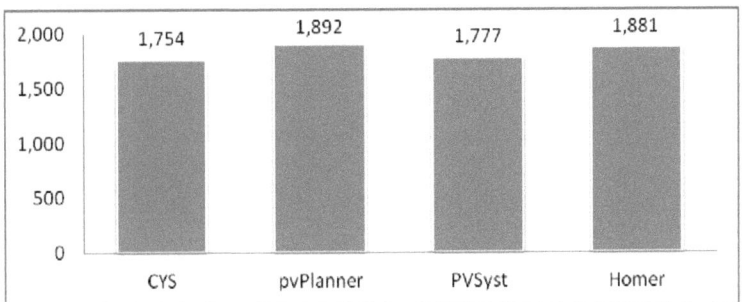

Figure 4.1 (a): Annual global horizontal radiation (kWh/m²)

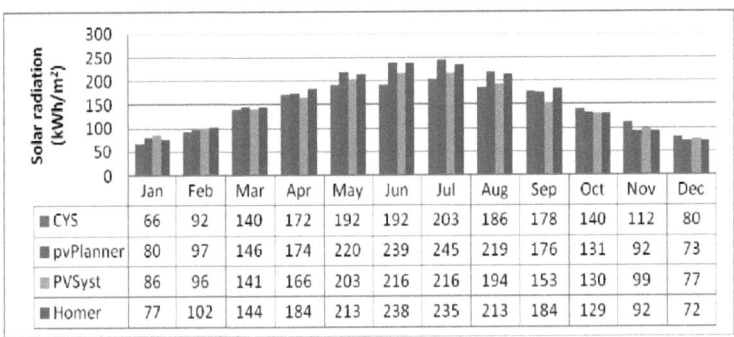

Figure 4.1 (b): Average monthly global horizontal radiation (kWh/m²)

Nicosia's measured annual average ambient temperature (20.4°C) for 2012 is a little higher than pvPlanner which modeled the highest average value among the other simulators, as presented in Figure 4.1 (c). PVsyst modeled the lowest average ambient temperature. HOMER does not model ambient temperature; the temperature modeling is accounted for in the derating factor (Geoffrey T. K et al, 2009). The derating factor is a scaling factor applied to the PV array power output to account for reduced output in real-world operating conditions compared to operating conditions at which the array was rated. All of these slight variations in primary data will affected the overall output results which will be discussed in this chapter. In Table 4.1, the measured solar radiation and ambient temperature of Nicosia for various seasons are summarized. As expected, the average ambient temperature and solar radiation are higest in summer season. The lowest values for both parameters are seen in the winter months.

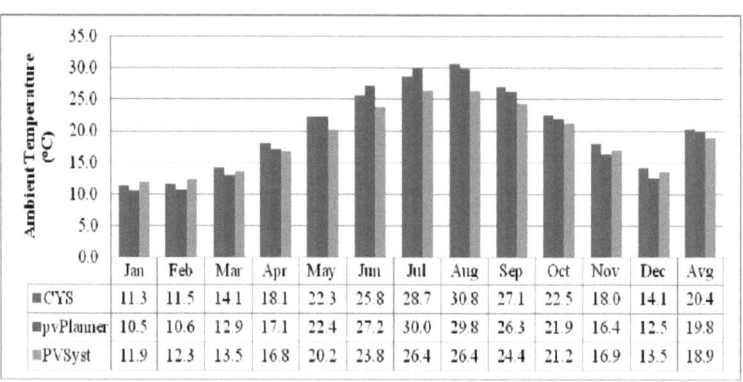

	Jan	Feb	Mar	Apr	May	Jun	Jul	Aug	Sep	Oct	Nov	Dec	Avg
■CYS	11.3	11.5	14.1	18.1	22.3	25.8	28.7	30.8	27.1	22.5	18.0	14.1	20.4
■pvPlanner	10.5	10.6	12.9	17.1	22.4	27.2	30.0	29.8	26.3	21.9	16.4	12.5	19.8
■PVSyst	11.9	12.3	13.5	16.8	20.2	23.8	26.4	26.4	24.4	21.2	16.9	13.5	18.9

Figure 4.1 (c): Average monthly ambient temperature (°C)

It can be clearly seen that the average ambient temperature during summer was of 28.4 °C which is 56.5%, 26.2%, 131% higher than average temperature during spring, autumn and winter, respectively, in Table 4.1.

Figure 4.1 (d) presents the solar duration (hours) for Nicosia. HOMER did not model daily or monthly solar duration but the annual solar duration was given as 4300 hours; this is 28% higher than the measured annual solar duration of Nicosia, which is 3159 hours at 2012. pvPlanner and PVsyst did not model daily or monthly solar duration. The average measured daily solar radiation for Nicosia is 8.8 hours, the maximum solar duration is 12.5 hours in July and minimum is 4.2 hours in January. It is found that during the summer period, Nicosia experiences longer solar radiation hours and shorter duration in winter.

Table 4.1: Seasonal ambient temperature and solar radiation data in Nicosia

		Temp	Average	Solar radiation	Average
		°C		(kWh/m^2)	
Spring	Mar	14.1		140	
Spring	Apr	18.1	18.2	172	168
Spring	May	22.3		192	
Summer	Jun	25.8		192	
Summer	Jul	28.7	28.4	203	194
Summer	Aug	30.8		186	
Autumn	Sep	27.1		178	
Autumn	Oct	22.5	22.5	140	143
Autumn	Nov	18.0		112	
Winter	Dec	14.1		80	
Winter	Jan	11.3	12.3	66	79
Winter	Feb	11.5		92	
	Annual	**20.4**		**1,754**	

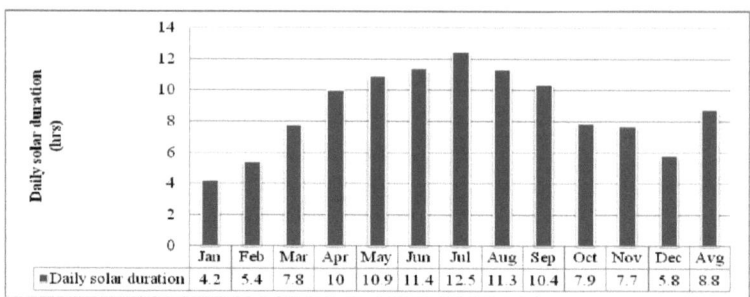

Figure 4.1 (d): Nicosia average daily solar duration (hours)

4.1 ENERGY PRODUCTION

The electrical energy production of a photovoltaic plant is an important measure of system performance. It is the amount of total alternating current (AC) energy delivered to the grid. The total and average energy productions in winter at CYS are 1,942 kWh and 637 kWh respectively. The highest energy production was witnessed during the summer season with total and average energy production of 4,102 kWh and 1,367 kWh respectively. The summer energy production was found to be 20% higher than spring production, 50% higher than the autumn production and 111% higher than the winter production. The energy production values are summarized in Table 4.2. The total annual energy production of the 5.76 kW PV system at CYS is 12,216 kWh; the system supplies 28% of its annual energy production in spring, 34% in summer, 22% in autumn and 16% in winter (Figure 4.2(a)).

44

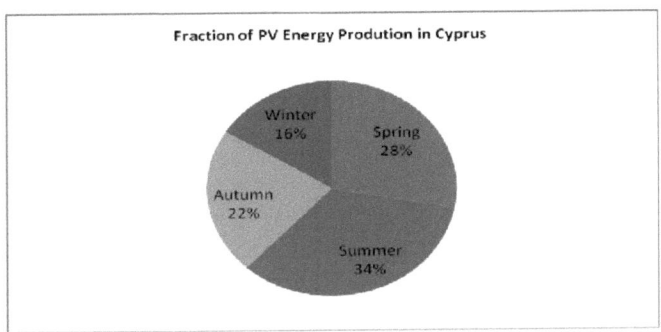

Figure 4.2 (a): Fraction of annual PV energy produced by 5.78 kW in Cyprus

When Table 4.1 and 4.2 are compared, it is seen that the lowest solar radiation and ambient temperature were obtained during the winter season. This accounted for the low energy production in this season. Also the highest solar radiation and duration was measured during the summer period, which is a main reason the highest energy production during this period. Temperature plays a major role on the performance of PV modules; when temperatures are high, PV modules performance drops (Swapnil D. et al., 2013). It was expected that due to the high temperature during the summer, there would be reduced energy production compared to spring and autumn season with temperatures of 18.2°C and 22.5°C respectively; but high solar radiation and longer solar durations during this period outweighs the negative effect of temperature on the plane of the array.

Also, the Standard Test Conditions air temperature of 25°C is 37.6%, 10.9%, and 103.3% higher than spring, autumn and winter average temperature respectively, while it is 12% less than summer temperature.

Table 4.2: AC energy production measured at CYS (kWh)

		AC Energy (kWh)		
		Monthly	**Total**	**Average**
Spring	Mar	1,044		
	Apr	1,167	3,431	1,144
	May	1,220		
Summer	Jun	1,387		
	Jul	1,400	4,102	1,367
	Aug	1,314		
Autumn	Sep	1,358		
	Oct	704	2,741	914
	Nov	679		
Winter	Dec	551		
	Jan	629	1,942	647
	Feb	762		
	Annual Total	**12,216**		

The annual average ambient temperature (20.4°C) at CYS is equivalent to the ambient temperature of the best case PV performance. According to the literature instead of the STC values, which means that a PV module will be typically rated at 25 °C under 1 kW/m^2, it is more important to consider the Nominal Operating Cell Temperatures (NOCT) of the PV module in order to determine the power output of the solar cell at the site. NOCT is defined as the temperature reached by open circuited cells in a module under irradiance on cell surface of 800 W/m^2, ambient temperature of 20°C, and wind Velocity of 1 m/s (Trinuruk P. et al. 2009). It is also observed that the ratio of total and average energy production in summer is more than twice the energy production in

46

winter. Interestingly, the ratios of summer radiation and summer temperature to winter radiation and winter temperature are 2.3:1 and 2.4:1 respectively. Similar trend is obtained when the summer energy production, solar radiation and ambient temperature is compared with the values obtained in spring and autumn seasons. As a result, it is observed that the ratio in the radiation is not directly reflected to the energy production. This is most probably due to higher ratios in the ambient temperature during these seasons also.

Table 4.3: Annual AC energy production of CYS and three simulators

	CYS	pvPlanner	PVsyst	HOMER
Annual energy production (kWh)	12,216	11,573	11,083	11,304
Fraction of CYS energy modeled by simulators	-	95%	91%	93%
Variance from actual production	-	5.3%	9.3%	7.5%

Table 4.3 compares the annual electrical energy production measured from CYS and the simulated energy from three PV software tools. The table and graphical representations comparing the detailed results of monthly energy production can be seen in Figures 4.2 and 4.3, and Table 4.4, respectively. pvPlanner simulated energy production with the least variation of 5.3% from the CYS annual energy production with a total production of 11,573 kWh, followed by HOMER with variation of 7.5% and total energy production of 11,304 kWh. It appears that all the simulators underestimated CYS annual energy production. The highest energy production deviation was observed in PVsyst. pvPlanner, PVsyst and HOMER modeled 95%, 93% and 91% of the annual energy production at CYS respectively.

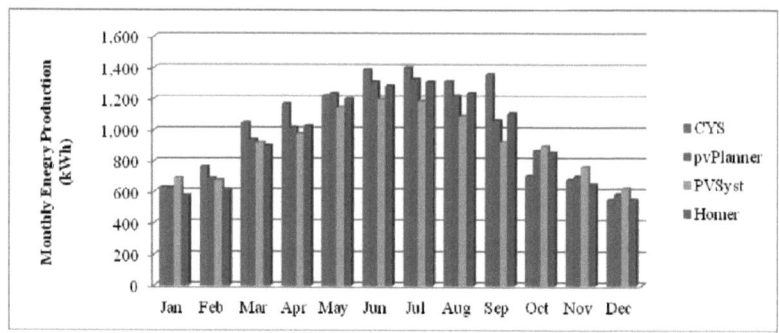

Figure 4.2 (b): Monthly energy production (kWh)

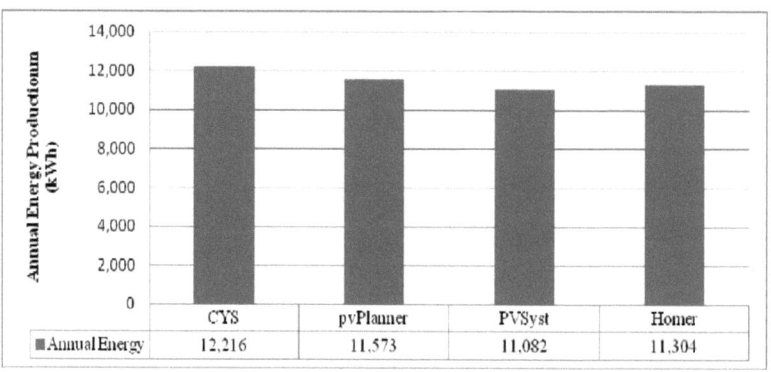

Figure 4.3: Annual energy production (kWh)

CYS monthly energy production did not match well with the monthly modeling results from all the simulation tools. The standard deviation of the three simulators' monthly energy production from the actual value is 10.7% for HOMER, 11.3% for pvPlanner and to 16.9% for PVsyst.

48

Table 4.4: Monthly AC Energy production of CYS and simulators (kWh)

Month	CYS	pvPlanner	PVsyst	HOMER
Jan	629	627	689	577
Feb	762	687	675	612
Mar	1,044	939	919	899
Apr	1,167	1,017	976	1,026
May	1,220	1,233	1,144	1,200
Jun	1,387	1,309	1,199	1,283
Jul	1,400	1,328	1,188	1,309
Aug	1,314	1,219	1,090	1,236
Sep	1,358	1,062	922	1,106
Oct	704	868	896	854
Nov	679	699	761	650
Dec	551	585	623	552
Annual Energy	**12,216**	**11,573**	**11,082**	**11,304**

When the average values were calculated, the results of the three PV simulation tools gave annual energy production amount to 11,320 kWh for 5.76 kW PV system under meteorological condition of Nicosia, NC. The measured value from CYS is 7.3% higher. It is also found that simulators monthly AC energy production is lower than actual CYS values except three months, October, November and December. In practice, it is acceptable that actual value of a PV system could be higher than simulated or predicted values. Generally, the conservative nature of some mathematical model and algorithm of the software tools is capable of underestimating or overestimating the performance of PV systems. In addition, the environmental and operational conditions of the real world system may not have been fully accounted for by the software tools. Briefly, solar radiation values at CYS have lower trend compared to the simulators but most probably, other meteorological data, such as solar duration values or ambient temperature values, have negative influence on the results of simulators. Use of long-term meteorological data by the simulators is a meaningful reason for this difference because CYS values are obtained just for one year; 2012. It is possible to have higher solar radiation but lower ambient temperature, clearness index and solar duration data with averaged long term data used by simulators compared to the measured data during 2012. Another possible reason for the variation in energy production could be the PV module peak capacity tolerance. Each module at CYS is rated 240W, and has a tolerance of 0/+5W. Therefore it is possible to obtain higher energy production values at higher solar radiation data during site applications. Since such may not have been accounted by the simulators; the fluctuations in the tolerance of the module might have led to daily or monthly increase in energy produced by CYS compared to the simulated energy by the software tools.

4.2 SPECIFIC YIELD

The brief definition of Specific Yield is; the division of the net energy output of PV system by the nameplate D.C power at Standard Test Condition of an installed PV array. Table 4.6 contains the monthly and season distribution of specific yield at CYS. The maximum yield of 243 kWh/kW_p (8.1 hours daily) was obtained in July during summer at CYS. The lowest yield value was 96 kWh/kW_p (3.2 hours daily) and obtained in December during winter. As expected it is seen that the normalized yield was higher in summer with average yield of 237 kWh/kW_p; while the lowest average season specific yield was 112 kWh/kW_p and was obtained during the winter. This implies that the PV system daily average peak operating hours in summer is 7.9 hours (7.9 kWh/kW_p) and 3.7 hours (3.7 kWh/kW_p) in winter.

Table 4.5: Monthly and seasonal specific yield at CYS

| | | Specific Yield (kWh/kWp) | |
		Monthly	Average
Spring	Mar	181	
	Apr	203	199
	May	212	
Summer	Jun	241	
	Jul	243	237
	Aug	228	
Autumn	Sep	236	
	Oct	122	159
	Nov	118	
Winter	Dec	96	
	Jan	109	112
	Feb	132	
	Annual Total	2,121	

From Figure 4.1 (d), the average daily solar duration in summer and winter are found to be 11.7 hours and 5.1 hours respectively. By simple arithmetic it is found that daily peak hour of operation in summer is about 67% of the solar duration and 72% in winter. The reason for the high percentage in winter can be traced to the effect of temperature in winter. In winter the ambient temperature is low even if the solar radiation is high. This enhances the performance of PV modules, hence it possible for peak operations with respect to available solar radiation and other factors. The specific yield is a derivation of the energy production; hence it is believed that all the factors responsible for the monthly and seasonal energy production and variations also affects the specific yield.

Figure 4.4 shows the monthly specific yield obtained from CYS and the PV simulators. The three PV simulation tools all have their highest specific yield (or highest monthly operating hours) in summer, particularly June and July; this corresponds to the period with highest solar radiation in the measured value at CYS. The period of lowest specific yield for the three simulators also match well with CYS measurement. Solar radiation is a major contributing factor to specific yield and so, all the PV generators, including CYS, operated at maximum and minimum duration at their peak rating during period of highest and lowest solar radiation respectively. The total annual specific yield values are shown in Figure 4.5. The annual CYS, pvPlanner, PVsyst and HOMER energy produced are 2121 kWh/kW$_p$, 2004 kWh/kW$_p$, 1924 kWh/kW$_p$ and 1963 kWh/kW$_p$ respectively. On the average, the specific yield for Nicosia will be predicted to be 1963 kWh/kW$_p$ by three simulation softwares. **Makrides G. et al (2010)** presented that installed fixed PV systems in Cyprus produced annual AC specific yields within the range of 1600–1700 kWh/kW$_p$. On the other hand, tracking systems has shown an AC energy yield of up to 2039 kWh/kW$_p$. **Pollikkas A. (2009)** also presented different specific yield values for fixed axis and two axis PV systems which vary between 1821 kWh/kW$_p$ and 2311 kWh/kW$_p$. These results also validates the measured results at one-axis tracking PV system at CYS and also results of three simulators.

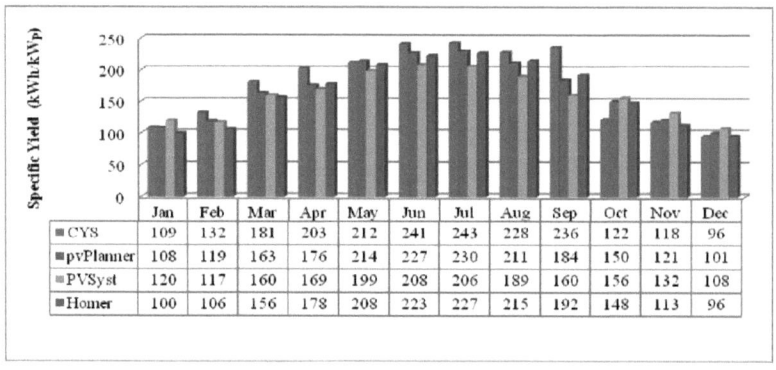

	Jan	Feb	Mar	Apr	May	Jun	Jul	Aug	Sep	Oct	Nov	Dec
CYS	109	132	181	203	212	241	243	228	236	122	118	96
pvPlanner	108	119	163	176	214	227	230	211	184	150	121	101
PVSyst	120	117	160	169	199	208	206	189	160	156	132	108
Homer	100	106	156	178	208	223	227	215	192	148	113	96

Figure 4.4: Monthly specific yield for system and PV simulation tools (kWh/kW$_p$)

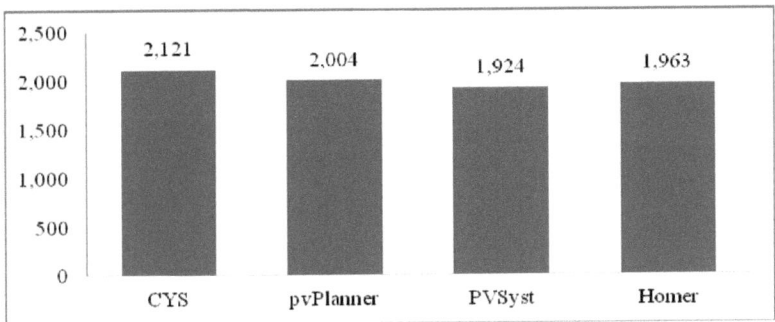

Figure 4.5: Annual specific yield (kWh/kW$_p$)

4.3 PERFORMANCE RATIO

The performance ratio is a measure of the quality of a PV plant that is independent of location and it is therefore often described as a quality factor (Marion B. et al., 2005). It does not quantify the amount of energy produced. When a PV system has low PR and located in a good irradiation site, it has a tendency to produce more energy than a PV system with high PR but located in a lower irradiation site (Cristian P. et al., 2009). The higher the PR the more solar energy is converted to electrical energy (R Faranda et al, 2011). Table 4.7 contains the performance ratio measured at CYS. The spring performance ratio is 84.9% while the summer performance ratio is 80.4%. It could be expected that the performance ratio in summer would be higher than the value measured in spring due to high solar radiation. Also, as mentioned earlier in this chapter, the energy production in summer is also higher than spring energy production.

Table 4.6: Performance ratio measured at CYS

		Performance Ratio		Solar Radiation	Avg	Temp	Avg	AC Energy	Avg
		Measured (%)		(kWh/m²)		°C		(kWh)	(kWh)
Spring	Mar	88.2		140		14.1		1,044	
	Apr	84.2	84.9	172	168	18.1	18.2	1,167	1,144
	May	82.3		192		22.3		1,220	
Summer	Jun	81.7		192		25.8		1,387	
	Jul	80.6	80.4	203	194	28.7	28.4	1,400	1,367
	Aug	78.9		186		30.8		1,314	
Autumn	Sep	78.1		178		27.1		1,358	
	Oct	77.9	77.9	140	143.4	22.5	22.5	704	913.6
	Nov	77.6		112		18.0		679	
Winter	Dec	75.0		80		14.1		551	
	Jan	82.0	79.8	66	79.4	11.3	12.3	629	647.4
	Feb	82.5		92		11.5		762	
	Annual	80.8		1,754		20		12,216	

This can be due to the negative influence of higher temperature in summer accounted for lower performance ratio during this period. The same trend can be seen in the performance ratio values of autumn and winter seasons. These results are confirmed by the study of Cristian p et al. (2009) that presented that performance ratio does not quantify the amount of energy produced. The annual performance ratio of CYS is 80.8%. At the time the measurements were taken, CYS systems had been installed for less than two years. According to National Renewable Energy Laboratory (NREL), such newly installed PV systems are expected to have minimum annual performance ratio of about 77%. Therefore, 80.8% is an acceptable PR value for CYS. Table 4.8 and Figure 4.6 compares CYS performance ratio with the simulation software tools.

Opposite to the results of PR values at CYS, during the summer period, all the PV simulators show considerable decrease in performance ratio except HOMER. Generally the variance in performance ratio from season to season is minimal for CYS and all the simulators. The average performance ratio from the simulation tools is 78.6%, this would represent the best case prediction for Nicosia, Cyprus.

Table 4.7: Performance ratio of CYS and PV simulators

	CYS	pvPlanner	PVsyst	HOMER
Jan	78.1	83.8	82.6	68.0
Feb	77.9	83.1	82.1	75.0
Mar	77.6	81.2	81.8	77.1
Apr	75.0	79.2	79.1	80.9
May	82.0	77.1	78.1	81.1
Jun	82.5	75.2	76.6	80.7
Jul	88.2	74.1	75.9	81.1
Aug	84.2	74.2	76.1	84.6
Sep	82.3	75.9	76.9	90.7
Oct	81.7	78.3	78.2	73.6
Nov	80.6	81.4	80.6	66.5
Dec	78.9	83.3	82.1	72.9
Year	**80.8**	**78.9**	**79.2**	**77.7**

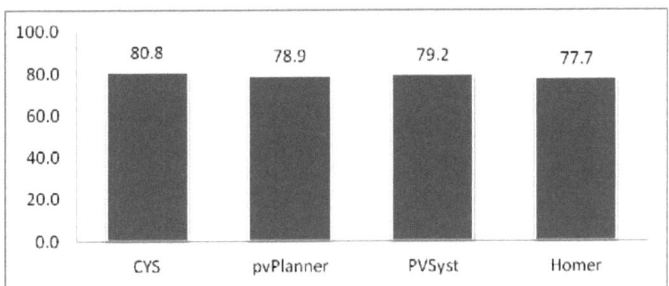

Figure 4.6: Annual PR of CYS and simulators

Table 4.8: Performance ratio of different PV systems in Cyprus

(Source: University of Cyprus, Nicosia)

System name	System size (kW$_p$)	Performance ratio (PR)
System 1 - Solon	1.54	79
System 2 - Sanyo	1.03	82
System 3 - Atersa	1.02	80
System 4 - Suntechnics	1.00	85
System 5 - Schott Solar	1.00	79
System 6 - BP Solar	1.11	74
System 7 - SolarWorld	0.99	79
System 8 - Schott Solar	1.02	80
System 9 - Würth	0.90	83
System 10 - First Solar	1.08	81
System 11 - Mitsubishi	1.00	80
System 12 - Schott Solar	0.97	77
System 13 - Atersa Tracker	1.02	81

A recent study on 13 photovoltaic systems conducted in University of Cyprus, Nicosia, as seen in Table 4.9, reveals that PR ranges from 74% - 85% under Cyprus meteorological conditions. It should be noted that all the systems represented in the Table are fixed and dual axis PV systems (Makrides G. et al. (2008 and 2010)). Therefore, measured results at CYS and simulation results look parallel with the finding in this reference.

According to NREL, it is emphasized that the standard performance ratio for a new PV system should be minimum 77%, and over time, the performance of the system is assumed to degrade. NREL has found that system performance will degrade 1% per year, which means that after 20 years, the system will be performing at 80% of its already suboptimal starting performance. For an average system, this means a performance ratio of just 61.6%. CYS is a new installation of less than two years operation and its annual PR is 80.8%. Therefore it is expected that in the next 20 years, the system performance ratio would be around 66.1%. According to the obtained results, all the simulators modeled performance ratios higher than NREL's recommendation.

4.4 CAPACITY FACTOR

The CF values represented in Figure 4.7 were calculated theoretically using the formula provided in previous chapters, except for HOMER which is the only PV simulation tool that modeled CF. CF only compares systems design and the ability of the systems to convert solar energy into electrical energy. It is the ratio of PV system DC energy output to rated energy output if the system operates throughout the year. Poullikkas A. (2009) presented south facing fixed (tilt angle 28°) and dual axis tracking PV systems at Cyprus with CF of 20.8% and 26% respectively. Makrides G. et al. (2008 and 2010) reported fixed PV system with CF of 19.4% and a dual axis PV system of CF of 23.3% at Cyprus.

To the knowledge of the authors of this study, there is no previous documentation or studies of the capacity factor for single axis PV systems in Cyprus. From the studies of Poullikkas A. and Markrides G. et al on fixed and dual axis PV systems, it is found that CF of a single axis PV system is expected to be between 19% and 26%. The single axis tracking PV system (tilt angle 30°) in this study has an annual CF of 25.06%. PVsyst, pvPlanner and HOMER modeled capacity factors of 22.73%, 23.74% and 23.19% respectively. A best case prediction for single axis PV systems in Nicosia would be average of the three simulators values, 23.22%.

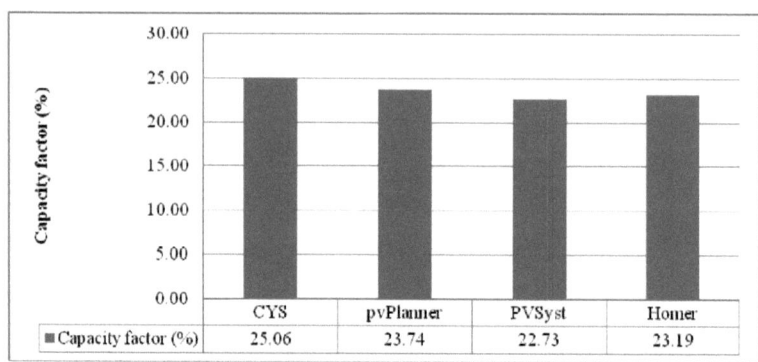

Figure 4.7: Capacity factors of CYS and three simulation tools

CONCLUSION

This study provides the evaluation of the performance of a grid tied, single -axis tracking 5.76 kW PV system located in Nicosia, the capital of NC. During this evaluation, four different parameters were considered; the energy production, specific yield, performance ratio and capacity factor. These data from the installed system were compared with simulated results obtained from three PV simulation software tools namely pvPlanner, PVsyst and HOMER. The meteorological condition of Nicosia is found to be suitable for PV systems operation. The solar radiation (an average global-in-plane radiation of 2565 kWh/m^2, 2580 kWh/m^2 and 2377 kWh/m^2 for CYS, pvPlanner and PVsyst, respectively) is parallel with the values obtained in the previous studies on Cyprus and Germany. The effect of the high temperature, over 28°C in summer period, and subsequent heat generated was found to have enormous negative effect on PV modules; it is, however, possible to minimize these negative effects by channeling the heat generated as a result of high temperature to other useful needs; such as photovoltaic-thermal (PVT) systems.

The annual energy production, specific yield, PR and CF of CYS were found to be 12,216 kWh, 2121 kWh/kW$_p$, 80.8%, 25.06%. These values are all higher than the corresponding values simulated by the PV software tools. A recent study of PV systems in Cyprus has found AC specific yield to be 2039 kWh/kW$_p$ and PR ranging from 73-85%. Another study found CF in Cyprus for single axis and dual axis PV systems to be 20.78 and 26.38%. All the modeling tools used in this study were found to be relatively accurate. In addition to the conservative nature of the simulators' algorithms, this difference in actual and simulated energy production can be traced to the use of meteorological data. CYS results were derived from actual and instant data during the operation of the PV system; while the simulators made use of historical data averages ranging from 10-20 years. The best case prediction for single axis PV systems in Nicosia as seen from the values obtained from the simulators are 11,320 kWh, 1963 kWh/kW$_p$, 78.6%, 23.22% for the energy production, specific yield, performance ratio and capacity factor respectively. pvPlanner, PVsyst and HOMER modeled 95%, 93% and 91% of the annual energy production at CYS respectively.

There are some uncertainties that may have affected the results obtained from CYS. Certain estimates and assumptions were made in the inputs to the modeling tools. The software tools respond to these assumptions differently depending on their algorithms and mathematical models. For instance, each one of the various derating factors considered by PVsyst were calculated based on a number of user inputs, whereas a single non temperature derating factor value was used as input into HOMER to encapsulate many possible derating factors (such as array mismatch, line resistive losses, panel soiling, shading, degradation, etc.). Also, pvPlanner could not accept module efficiency, number and rating of individual modules as input. The arrangement (series or parallel) of modules in pvPlanner and HOMER were not explicitly documented by the simulators; hence we assumed they were connected in series. The estimates and assumptions of input into the PV software simulation tools are dependent on the users and their experience. In addition, some of the software tools do not allow manual input of some data, especially for solar radiation as in pvPlanner. Among the three simulators used in this study, only PVsyst allow for the flexibility of selection of various user inputs, thereby increasing the reliability of its results. Generally, a more reliable and fair comparison would have been possible if all the data (solar radiation, ambient temperature, humidity, clearness index etc.) measured at the system site were entered as inputs to the simulation tools.

As a result, it is clearly observed that, installation of PV systems in Cyprus is feasible with very high performance results according to the evaluation of a real world application in Nicosia. Also, it is found that the three PV simulation software's results were slightly different from the installed system but enough to give right idea of the performance of a PV system in NC. A number of further studies can be conducted to build on the results of this study, such as:

- Hybrid PVT systems
- Optimization of tilt angles
- Effect of ambient temperature variation
- Use of other PV technologies.
- Investigating the sensitivity of the models to different input parameters and conditions.

REFERENCES

Abbasoglu S. et al.
2010.   Energy Trend and Energy Efficiency in Turkish republic of North
                       Cyprus. 5th International Ege Symposium and exhortation
                       (IEESE-5) 27-30

Amita C. et al.
2013.   *Optimization of solar by varying tilt angle/slope.* International
                       Journal for Emerging Technology and Advanced Engineering,
                       Volume 3, Issue 4

Boyle, G.
2004.   *Renewable Energy: Power for a sustainable future.* New York:
                       Oxford University Press

Cristian P. et al.
2009.   Acta Technica Corviniensis Bulletin Of Engineering, Scientific
                       Supplement Of Annals Of Faculty Engineering Hunedoara
                       International Journal Of Engineering. Issn: 1584-2665 – Print, Issn:
                       1584-2673 - Cd-Rom

Del C.J. and Von R.B.
1999.   *Temperature-induced changes in the performance of amorphous
                       silicon multi-junction modules in controlled light-soaking.* Progress
                       in Photovoltaics 7, 101–112.

Dunlop, J.P.
2010.   *Photovoltaic Systems.* Illinois: American Technical Publishers

Geoffrey T. K. et al.
2009.   *Models Used to Assess the Performance of Photovoltaic Systems*,
                       Sandia National Laboratories, US Department of Energy. SAND
                       2009-8258

Gottschalg, R. et al
2004.   *A critical appraisal of the factors affecting energy production
                       from amorphous silicon photovoltaic arrays in a maritime
                       climate.* Solar Energy 77, 909–916.

Gottschalg, R. et al.
2005.   *The effect of spectral variations on the performance parameters of
                       single and double junction amorphous silicon solar cells.* Solar
                       Energy Materials and Solar Cells 85, 415–428.

Griffith J. S. et al.

1981.   *Some tests of flat plate photovoltaic module cell temperatures in simulated field conditions.* Proc. 15th IEEE Photovoltaic Specialists Conference, Kissimmee, FL, 1981; p.822-30

Huld T. et al.

1981.   *Mapping the performance of PV modules, effects of module types and data averaging.* Science direct: Solar energy 84 (2010) 324 - 338

Lazar R.

2013.   *A comparison of solar panel efficiency: A definition and calculation of PV efficiency.* Facts about solar energy website. URL: http://www.factsaboutsolar energy.us/solar-panel-efficiency.html (retrieved October 14, 2014)

Makrides G. et al.

2008.   *Performance assessment of different photovoltaic systems under identical field conditions of high irradiance.* In: 4th photovoltaic science application and technology conference; p. 199–202

Makrides G. et al.

2010.   *Potential of Photovoltaic systems in countries with high solar radiation.* Renewable and Sustainable Energy Reviews 14 (2010) 754–762

Makrides G. et al.

         Performance assessment of different photovoltaic systems under identical field conditions of high irradiance. Department of Electrical and Computer Engineering, University of Cyprus, Nicosia

Marion B. et al.

2005.   *Performance Parameters for Grid-connected PV System.* In: 31st IEEE Photovoltaic Specialist Conference, Lake Buena Vista FL

Martin, N. and Ruiz, J.M.

2001   *Calculation of the PV modules angular losses under field conditions by means of an analytical model.* Solar Energy Materials & Solar Cells 70, 25–38.

Nikolaeva-Dimitrova M. et al.

2008.   *Controlled Conditioning of a-Si:H thin film modules for efficiency prediction.* Thin Solid Films 516, 6902–6906.

Paul Gilman

2004.   Homer energy website: www.homerenergy.com (Retrieved October 13, 2013)

PVsyst website

         http://files.pvsyst.com/pvsyst5.pdf (Retrieved October 13, 2013)

Ralf M.
2009.   *Maintaining the Performance Ratio of PV Systems*, Solar Industry
               Magazine, Solar Feeds website. URL:
               http://www.solarfeeds.com/maintaining-the-performance-ratio-of-pv-
               systems/ (Retrieved October 20, 2013)

Roney, J. M.
2013.   World Solar *Power Topped 100, 000 megawatts in 2012*. Earth
               Policy Website. URL: http://www.earth-policy.org/indicators
               /C47/solar_power_2013 (Retrieved September 28, 2013).

Sharma V. et al.
2013.   *Performance assessment of different photovoltaic technologies
               under similar outdoor conditions*. Solar direct: Energy 58 (2013)
               511 – 518

SolarGIS website
               http://solargis.info/doc/4 (Retrieved October 13, 2013)

Stephen, L.
2013.   *2/3rds of Global Solar PV has been installed in the last 2.5 years*.
               Green Tech Media Website. URL: http://www.greentechmedia
               .com/articles/read/chart-2-3rds-of-global-solar-pv-has-been-
               connected-in-the-last-2.5-years. (Retrieved September 28, 2013)

Swapnil D. et al.
2013.   Temperature Dependent Photovoltaic (PV) Efficiency and Its
               Effect on PV Production in the World - A Review. PV Asia
               Pacific Conference 2012 Energy Procedia 33 (2013) 311 – 321

Trinuruk P. et al.
2009.   Estimating operating cell temperature of BIPV modules in
               Thailand, Renewable Energy 34 (2009) 2515–2523

Zinsser B. et al.
2007.   *Annual energy yield of 13 photovoltaic technologies in Germany and
               Cyprus*. In: Proceedings of the 22nd European photovoltaic solar
               energy conference; 2007. p. 3114–7.

Anonymous
2009.   *Falling panel prices could bring solar close to grid parity*. URL:
               http://greenecon.net/falling-panel-prices-could-bring-solar-closer-to-
               grid-parity/energy_economics.html (Retrieved September 28, 2013).

Anonymous

Types of solar photovoltaic mounting systems: 5 common mounting systems. Atlantech Solar Website. URL: http:// atlantechsolar.com/photovoltaic_solar _mounting_systems.html (retrieved October 1, 2013).

Anonymous

U.S. Department of Energy, Energy efficiency and Renewable Energy, Energy Basics. URL:http://www.eere.energy.gov/basics/ renewable_energy/pv_system_performance. html (Retrieved September 30, 2013).

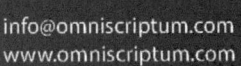